Snowflake Chronic

4

CLIMATE HYSTERIA

Mark Lawson

Connor Court Publishing

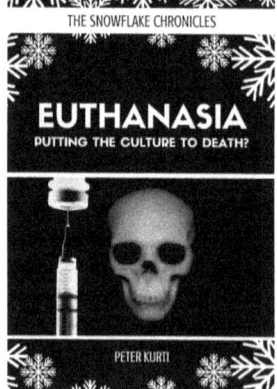

Snowflake Chronicles
1
*Right thinking on
Abortion*

Nicola Wright

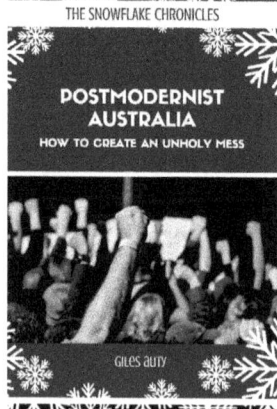

Snowflake Chronicles
2
*Postmodernist
Australia*

Giles Auty

Snowflake Chronicles
3
*Euthanaisa: Putting the
Culture to Death*

Peter Kurti

The Snowflake Chronicles

A word from the publisher

The Snowflake Chronicles book series is an initiative dedicated to pushing the limits of political correctness.

The literary world has become the bastion of progressive thinking that limits free speech and limits topics that were once the norm. The first book of the series was on Abortion. Try talking about Abortion, from a pro-life perspective in academia or even on campuses. It is unthinkable. How many publishers actively publish books that give a pro-life perspective? Are they allowed to? Such is the state of the debate that some topics are purely off-limit, even in a discussion, especially in the world of publishing. So it is in this void that the Snowflake Chronicles were born. We are not snowflakes, we will not crumble or dissolve by listening to an argument. Topics like Abortion, Euthanasia, Coal, Climate Change, Marriage, the minimum wage, immigration, and culture are not topics where the debate has moved on, leaving us behind. We want to bring these topics back from their extinction and show that these so-called contrarian views have merit and will be debated, promoted and encouraged.

So if you are a snowflake, please use these books as a therapy for your recovery. These books will hopefully enlighten the so-called progressive thinker and in some cases cure them. For those who agree with the sentiments of the books, you may find them inspiring and entertaining. Can we encourage you, after you, read them, to help save the world by passing the book onto a snowflake.

Snowflake

A very sensitive person. Someone who is easily hurt or offended by the statements or actions of others.

Published in 2019 by Connor Court Publishing Pty Ltd

Connor Court Publishing Pty Ltd
PO Box 7257
Redland Bay QLD 4165

sales@connorcourt.com
www.connorcourtpublishing.com.au

Phone 0497 900 685

ISBN: 978 1925826418

Cover arrangement Nicola Wright

Printed in Australia

contrarian

adjective
1. opposing or rejecting popular opinion or current practice.

1
Hype, hope and hyperbole

The Red Queen shook her head, "You may call it nonsense if you like," she said, "but I've heard nonsense, compared with which that would be as sensible as a dictionary!"

- Alice Through the Looking Glass – Lewis Carroll

Once a week in a conference room in Salt Lake City, in the American state of Utah, a dozen people gather for a session of climate change grief counselling. Convened by a Laura Schmidt, a full-time activist with a masters in environmental humanities, the sessions permit the participants to vent over all the things not being done about climate change, and how the participants themselves are doing things that contribute to a problem (such as drive cars to the meetings, we suppose) that they imagine will affect their loved ones.

One tale to emerge from these fraught sessions is that of a woman who, when confronted by piles of merchandise in a store produced and packaged in all sorts of energy intensive ways, had to retreat to her car to recover for a time before she could face shopping again. Ms Schmidt, who organised additional sessions following the election of Donald Trump as American president (Trump easily won Utah, a traditionally Republican state), has developed a ten-step coping program loosely based on the Alcoholics Anonymous program to cope with this sort of stress. Right.

This is an extreme but hardly surprising indication of the hysteria about global warming/climate change amongst those predisposed to believe in catastrophes – hysteria largely unconnected with anything actually happening in the earth's climate. Scientists and activists have been screaming that the end is nigh and thumping the alarm button on global warming for all they are worth for more than thirty years, as of early 2019. In that time, any number of climate tipping points, deadlines for action, and projected climate disasters have come and gone without anything much happening. Snow is supposed to have vanished, the Arctic, Antarctic and Greenland ice sheets long melted, Bangladesh is meant to have disappeared beneath the waves along with the assorted tropical islands, and Adelaide should be deserted due to lack of drinking water.

Like all veteran forecasters, however, and in the grand tradition of political pundits who were almost all totally wrong over the election of Donald Trump as president, the many global warming activists and scientist-activists have brushed off their abysmal forecasting record and kept on talking. Climate doom is just around the corner, and anyone who dares to question this is obviously in the pay of big energy (I only wish) or suffering from a psychological disorder.

One result of all this talk, supported by a great deal of government funding, is that the world of climate alarmism has split from the real world. This alarmist world has its own logic, and is on the verge of climate doom, and never mind what's happening in the real world. For example, what is happening in Salt Lake City that causes grief worthy of counselling? Have temperatures become markedly higher? As temperatures in that region vary wildly between hot summers and chilly winters, individuals would have trouble noticing any change. Has rainfall been reduced? Again no, but then the area has been classified as semi-arid (the classification is in dispute) so it may be difficult to tell. Salt Lake City is a long way inland so sea level increases should not be a problem, even if sea levels happen to be increasing very much, which they are not.

The city does have an air quality problem, however. An air inversion caused by the surrounding mountains traps smog, including exhaust from a lot of vehicles. It is one of the last American cities to have such a problem. This has nothing to do with climate change or with carbon dioxide, which is odourless, colourless and spreads around the globe quickly. Smog involves the likes of sulphates and aerosols which might also be produced in the same sort of emissions as CO2, but can be dealt with locally through strong measures. However, the sort of people who go to climate grief counselling are not about to make the distinction between carbon dioxide and pollutants. As far as they are concerned, local pollution is the same as global CO2 concentrations and we are all doomed, unless industry stops.

The climate grief, or perhaps climate silliness industry is by no means confined to Utah. In March 2017 the American Psychological Association, in conjunction with the organisations Climate for Health and ecoAmerica produced a report entitled *Mental Health and Our Changing Climate – Impact, Implications and Guidance.* This points to all sorts of mental health effects that will result from climate change, which includes "natural disasters exacerbated by climate change, like floods, storms, wildfires, and heatwaves. Other effects surface more gradually from changing temperatures and rising sea levels that cause forced migration. Weakened infrastructure and less secure food systems are examples of indirect climate impacts on society's physical and mental health".

Further on in the report we are told: "Climate change–induced extreme weather, changing weather patterns, damaged food and water resources, and polluted air impact human mental health. Increased levels of stress and distress from these factors can also put strains on social relationships and even have impacts on physical health, such as memory loss, sleep disorders, immune suppression, and changes in digestion."

Digestion? Stomach upsets? So that's my problem. More seriously, the hysteria of this report is evident by its references to forced migrations caused by rising sea levels, and less secure food systems, when both of those scare stories have already proved to be a complete bust, as is noted elsewhere in this essay. In addition, if the Intergovernmental Panel on Climate Change has ever claimed any link between fires and climate change this claim is difficult to find. Although the IPCC has previously claimed strong links between climate change and storms these claims were drastically toned down in the panel's 2014 report (the latest) – a point also explored elsewhere in the essay. Those are just a few of the problems with the APA's no-doubt well-meant publication. People who are flooded out or affected by storms would certainly be subject to stress, but have they been subjected to that additional stress due to a change in climate, or due to a decision to move near the coast or onto a flood plain?

This insistence on a crisis in a Utah and around the world by a segment of the population that is also naturally vocal and predominate among journalists has prompted governments into thinking that they should do something. If only to be seen to be doing something because of all the screaming. The trouble is that at the same time those governments do not want to annoy the voters with strong measures. This has had odd results.

Before the fall of the Berlin Wall when the inefficiency of the old Eastern-bloc regimes was a byword, workers in those countries use to joke that they pretended to work, and their governments pretended to pay them. That is about where we are in the parallel world of the climate debate. Scientists are pretending that there is a crisis while governments are pretending to do something about it.

Perhaps pretending is a harsh word for many of the scientists engaged in this issue. Just as the governments of the Soviet bloc tried to do the right thing by their workers, provided they did not want real social change, many scientists are genuinely concerned about the future. Any number of failed predictions are not going to shake this certainly, particularly as they now have decades of their careers invested in the theory.

Instead we are subjected to a lot of hype that borders on propaganda reminiscent of the old Soviet bloc regimes. Australia's admittedly very hot summer of early 2017 was a handy illustration of what journalists and academics had been insisting on all along, that climate doom was upon us and it was going to get worse. They did not trouble to mention that the winter that hit Europe and North America at the same time was bitterly cold, or place the summer heat in context. How hot had the previous hottest summer been and when was it? That would confuse the message. When that winter and the even colder northern hemisphere winter of 2017-18 had to be mentioned, they were blamed on global warming.

A press release issued by Australia's Bureau of Meteorology in early January 2019 continued this grand tradition by announcing that 2018 was the third warmest year on record, as part of a succession of hot years. This is undoubtedly true, for reasons to be explored in the next chapter, but just how did temperatures of those years compare with temperatures of, say, the late 1930s and early 1940s when temperatures in Australia were also known to be high? How much really has changed? The same newspapers that trumpeted the BoM's release, incidentally, would have carried news of bitterly cold conditions in Europe, including snow on some of the Greek Islands and thirteen deaths mainly due to avalanches in one week.

As for pretending to do something about the crisis, the Paris agreement of late 2015 was arguably little more than a show document, even before Donald Trump, elected president with the avowed intention of repudiating the US part of the treaty, had withdrawn from the treaty as he promised. This he can do as it was never a treaty from the American point of view. It was a presidential agreement. But the Paris meeting was only one of an endless succession of international meetings, where tens of thousands of government delegates and representatives from Non-Government Organisations have flown to exotic locations on emissions-generating jets to discuss the reduction of emissions. These meetings have largely been a waste of time, despite trumpeting of agreements by activists, because the task they are trying to accomplish is simply too large.

Getting governments to stop using nerve gas on their own citizens, say, or stop supporting terrorism, or modify stringent bank secrecy rules which help criminals to launder money is bad enough, but insisting on global rules that might crimp industry and lose jobs when the economy is in a down turn and an election around the corner, is in another category. As a result, in this alternate world of climate change action, governments will claim that they are concerned about the environment and the next day issue more free emissions permits to industries to win votes. Climate activists reinforce this behaviour by applauding governments who talk big about reducing emissions, particularly at the endless international meetings, but then fail to speak out when those same governments ignore their own hype. The Chinese government gets a particularly easy run by activists in this regard. Half the world's coal is dug up in China, but activists will regularly laud the country as an environmental leader working to kill off the coal industry.

This hype is regularly seasoned with flights of fantasies about green energy which are just as nonsensical as anything heard by the Red Queen in the quote at the beginning of this chapter. The world's electricity generators, we are regularly told, will be swept aside to be replaced by clean and green photovoltaic panels and wind generators. Add in a few renewable base load generators – never mind that these have not been invented yet, despite years of trying - and the coal industry will have to close. Doubters are shown computer models compiled by academics with impressive credentials which shows that a 100 per cent renewables electricity network is possible at minimal cost.

This book will not descend into the rabbit hole of renewable energy, that is a book in itself. Instead this essay will mainly emphasise the many bung forecasts about climate. Any discussion of the science by me would be easy to dismiss as I am not qualified in the area. Instead, it will show that no matter what you may think of the science, it is unsafe to use the forecasts for public policy. This is quite separate from the issue of whether the community choses to do anything about emissions for moral reasons.

This essay may well achieve little. Attempts to correct any part of the vast amount of self-evident nonsense spoken and written every day on climate are akin to shouting into a hurricane. For every letter to a newspaper that, say, counters a report involving any of the above, a host of indignant replies appear the next day. Assertions that the economy will be wrecked, that diseases will spread, storms will grow worse and seas will rise – claims that the IPCC have now either abandoned or toned down - are all reasserted. A good disaster story, after all, is very appealing.

This essay points to some of the problems with the more outrageous claims made by the hundreds of thousands of climate change proponents every day and, sadly, that will have to do.

2

Foggy forecasting

Prediction is very difficult, especially if it's about the future.

- Nils Bohr, Nobel Laureate in Physics

I said this essay would not be about the science and it isn't. To briefly explain my own position, as a layman with basic scientific training, after going over the evidence I strongly suspect that what scientists have been measuring and wringing their collective hands over, may well be entirely natural. I'm not what is known as a "luke warmer", as those skeptics who believe human activity may be causing some warming describe themselves. I am a no-warmer, I suppose. But to explain why I suspect this would take too long and bore most of those who might read this book and, as noted, can be dismissed with a wave by those who do not wish to hear the arguments. Instead, I've taken the different approach of checking a few of the basic forecasts made by the climate change industry against reality. Let us start with temperatures.

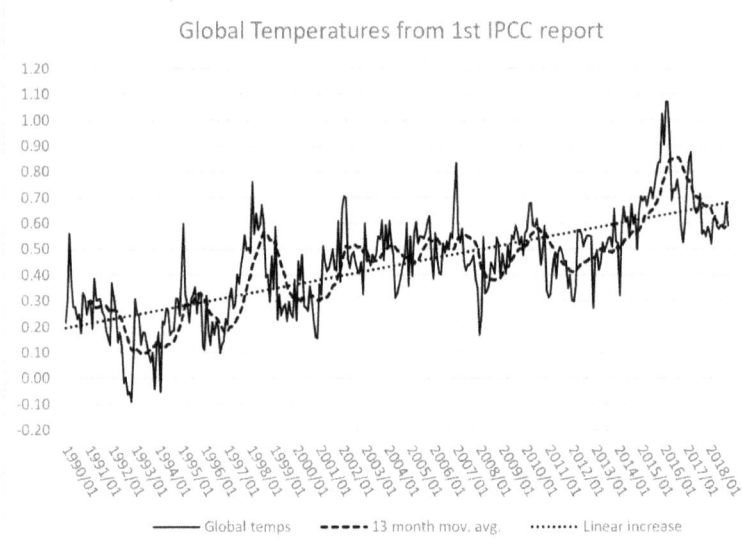

Global Temperatures from 1st IPCC report

Source: This is a graph of the monthly changes in global temperatures taken direct from the UK based Hadley Climate Science Unit.

Readers will note that I started from 1990. Why that year? Professor James Hansen, the patron saint of climate change warnings, first fronted a congressional committee with his concerns in 1988, thirty one years ago. The Charney report, which sets out the basics of global warming theory, was released in the 1970s. However, the era of hand wringing over forecasted temperature changes using massive computer models to forecast our climate future really kicked off when the International Panel on Climate Change issued the first of several climate forecasts in 1990. These reports can be accessed on the IPCC site and so checked against actual results.

The 1990 report gives three scenarios of which the most alarming is the Business As Usual case. This states that the minimum expected temperature increase will be 0.2 degrees, or two tenths of a degree centigrade, a decade. That should work out to a total temperature increase of about 0.6 or six tenths of a degree centigrade, give or take, since the report. The IPCC gave a mid-range or best guess forecast of 0.3 degrees a decade and an upper range of half a degree a decade. Subsequent IPCC reports issued in 1995, 2001, 2007 and 2014 made broadly similar forecasts.

Several groups compile temperature graphs of the earth, from both land stations and from satellites, of which perhaps the best known is the temperature series compiled by the Hadley Climate Centre in the UK. I have downloaded the centre's standard time series, available from the site, and added a moving 13-month moving average trend line plus an average straight-line increase courtesy of standard graphing program Excel. As readers can see from the straight-line increase, between 1990 to early-2019, a period of 29 years, it is possible to argue that earth's global temperatures increased by somewhere under half a degree, give or take. That works out to an average increase of maybe 0.17 degrees per decade, again give or take. This is below the bottom end of the IPCC forecast made in 1990 and a world away from the mid-range and top end forecasts.

Global warming activists, however, will pounce on this result triumphantly, waving away the problem that the result is far milder than the forecast. There you are, there has been no pause. Temperatures are increasing. The skeptics are wrong.

Sure, the world has warmed. The problem, as climate change proponents will carefully avoid mentioning, is that the earth's climate is known to have varied very substantially in the past including into and out of ice ages and mini ice-ages (a period of colder temperatures), including one that gripped Europe for decades up to the mid-nineteenth century. Scientists have been investigating past climate states for decades by looking at ice cores in the Antarctic, rock formations in caves in New Zealand, samples of pollen from dirt in Peru, and evidence of where the tree line stopped in the alps in medieval Europe (as an indication of temperatures) to construct temperature charts of past eons. From all of that work and extensive analysis of modern instrument readings can we say that the warming of the past 30 years is unusual when compared to past periods of warming and cooling?

The answer is that we can't, or at least scientists have yet to present any definite analysis that I have seen that the warming of the period is usual or different from known past changes. A closer look at the graph points to further problems for the large and active global warming industry. For the bulk of the warming we are looking at in my modest graph occurs between 1990 and 2002. Then between that year and 2015 or so nothing much seems to happen at all, despite all the additional CO2 supposedly pouring into the atmosphere from industrial activity. This is the famous climate "pause" which has been the subject of much argument. A careful re-analysis of the data by highly trained climate scientists, we are told, eliminates this pause. Others have then claimed that the re-analysis is flawed.

Instead of spending a lot of time on claim and counter claim in this area lets skip to the end where temperatures were pushed up quite sharply by a very strong El Niño effect. As is well known, or should be well known, the El Niño effect is a climate cycle that occurs at irregular intervals. Its effects/symptoms include generally reduced rainfall along Australia's eastern seaboard and higher global temperatures. The El Niño of 2015-2016 was a particularly strong one, permitting global warming activists to claim, possibly truthfully for once, that those two years were the hottest on record. There you are, global warming is real, it just took 30 years for anything to happen.

But if the El Niño affect was a strong one, wouldn't that affect the overall, average increase we can see on the chart – that is, push up the straight line? You bet, and an unintentional ally in adjusting the results for that effect is the World Meteorological Association. An extraordinary WMO press release which owes more to activism than science based on the organisation's own report, issued in March, 2017, says:

"Global temperatures continue to be consistent with a warming trend of 0.1 °C to 0.2 °C per decade, according

15

to the WMO report…

…The powerful 2015/2016 El Niño event boosted warming in 2016, on top of long-term climate change caused by greenhouse gas emissions."

The WMO press release writers do not seem to be aware that the estimate of 0.1-0.2 degrees per decade is actually right on the bottom or seriously below not only the minimum projections given in the IPCC 1990 report, but the reports of 1995, 2001, 2007 and 2014. Faced with these problems the climate establishment has reluctantly admitted that the climate models used to forecast the big temperature increases are not perfect.

A June 19, 2017 paper in *Nature Geoscience*, entitled *Causes of differences in model and satellite tropospheric warming rates*, states in part: 'We conclude that model overestimation of tropospheric (upper atmosphere) warming in the early 21st century is partly due to systematic deficiencies in the post-2000 external forcings used in model simulations.'

This basically means that the climate models gave the wrong results because they did not take into account all the factors affecting climate. Scientists have made this observation before and been bitterly attacked for their troubles, but this paper is notable for including, as authors, the global warming leading light Professor Michael Mann, of Pennsylvania State University and one of Australia's most distinguished scientists in this area, Professor Matthew England of the University of NSW.

The authors blame the difference between observation and reality on a combination of 'internal variations' and short-term natural cooling such as volcanic eruptions injecting material into the atmosphere. In other words they are not abandoning global warming as a theory, merely explaining why the models are falling short by adding other factors. The implication is that the models will still accurately forecast warming over the long term. Repent of your sinful ways of consumption and energy use or judgement will be upon you.

We won't challenge this paper here, only note that for whatever reason the actual results fell short of the minimum forecast. In fact, they are a long way below the doom and gloom constantly foretold by climate prophets.

As all attempts to forecast increases since the birth of organised climate forecasting in 1990 have proved wide of the mark, as have rainfall forecasts and much else besides (points to be explored elsewhere in the essay), it is difficult to take yet more forecasts of doom and gloom seriously, or base any policy to curb emissions on them. The WMO report cited before pointed to a host of bad weather events, as proof that the planet is being ruined by extra carbon dioxide. Well, okay, bad stuff happened. Can any of this bad stuff be forecast by the theory, or is the bad stuff confirmation of the theory because it is bad? For that matter what was unusual about the bad stuff? Again, this is far from convincing.

Activists (who may also be scientists) unaware of the

grudging admission in the *Geoscience* paper, incidentally, may still respond to complaints that temperatures don't seem to be moving very much by stating that temperature increases are within forecast range. A request for further and better particulars will result in the activist grandly pointing to a graph where the line showing the earth's temperature, usually starting from 2000, is between two other, widely separated lines – the upper and lower forecast estimates.

Anyone familiar with the sorry history of forecasting will be completely unimpressed. The one and only test of any system used to make forecasts is whether it can successfully forecast results unknown at the time the forecast was made. Instead of rigorously assessing past forecasts, however, scientists have tweaked updated versions of existing models to follow current climate conditions, to then claim that the forecasts of those same models for climate conditions decades into the future are all but gospel.

A closer look at these forecasting systems reveal more problems. The temperature models require settings in poorly understood area such as cloud cover and repeated runs to get a result. But those models in turn rely on forecasts of CO_2 concentrations in the air. Those concentration forecasts, in turn, rely on forecasts for population growth and economic and technology (for industrial emissions). At the other end of the process, forecasts of economic damage resulting from temperature increases rely on little known economic techniques, and on a discount value for estimating damage in modern day dollar values. Economists have been arguing for years over what discount rate should be used.

All clear? To examine all of that in detail this essay would have to be many times its present length, but in my view all these interacting forecasting systems have major problems. The estimates of future damage due to global warming are, in particular, little more than guesses based on fantasy. Public policy cannot be based on this nonsense.

This nonsense quickly degenerates into farce when commentators, including some far less qualified than myself, stray into areas well away from temperatures while still trying to claim the same forecasting status attributed to temperature projections. Read on.

3
Rain, snow and ice

"If we don't get three inches, man,

Or four to break this drought,

We'll all be rooned," said Hanrahan,

"Before the year is out."

- Said Hanrahan, John O'Brien

My favourite story about global warming forecasting is about the *Sydney Morning Herald* journalist who went to the Antarctic to report first hand on how global warming was affecting the continent, only for her ship to get stuck in ice.

Nicky Phillips, the Herald's science editor of the time (2014), was travelling to the Antarctic along with another journalist, Colin Cosier, in the Russian-flagged Akademik Shokalskiy as guests of the Australian Antarctic Division's media program. Phillips and

Cosier were part of a larger expedition which included scientists investigating how climate change was devastating the Antarctic and its "fragile" eco-system (if an eco-system survives in the Antarctic, how can it be fragile?)

Phillips wrote various articles before going about how she intended to use the trip to report on the changes caused by global warming. Nature declined to cooperate. The Shokalskiy became stuck in ice well before reaching the Antarctic. Another ship set to get close enough to the Shokalskiy to extract the journalists and other passengers by helicopter, the Chinese icebreaker the Xue Long, also became stuck in ice. A third ship was able to rescue the stranded passengers through an elaborate evacuation plan that involved them walking over pack ice.

On her return to Sydney, Phillips wrote an article stating that her ship getting stuck in ice was really due to global warming. Without troubling ourselves with the details of Phillips argument, there was another theory doing the rounds at the time that the abnormally cold North American winter of that year was also really due to global warming. This was along the lines of warming oceans melting ice in the arctic which released warm air. That warm air, or perhaps warmer air in the context of the arctic, disturbed the pocket of cold air which sits above the arctic, pushing it South. A variation on this mechanism was also used to explain the freezing Northern winter of 2017-18. The question of whether this explanation is correct – and there may well be some truth in it - is beside the point. What it seems to mean is that a major effect of global warming is a succession of arctic-cold Northern hemisphere winters. This does not make much sense to me. The plight of the Akademik Shokalskiy attracted plenty of international media attention, incidentally, without any of the items mentioning that it was carrying scientists studying global warming.

Phillips' expedition is not the only one to check on global warming only to be defeated by icy conditions. In the Northern winter of 2017, a multi-year multi-million dollar study of climate change in the Arctic lead by scientists from the Canadian university of Manitoba was postponed for a year due to extreme ice conditions around Newfoundland and Lapland. The icebreaker they were to use was needed to rescue fishermen trapped in the ice and, in any case, had trouble getting through the thicker than normal ice. This thicker ice was explained away as ice from further north that had broken off and floated south due to climate change.

In 2010 the UK Caitlin expedition, backed by the Prince of Wales, sent to assess sea ice thickness in the Arctic ocean made it just half the distance to the North pole due to extreme conditions. In August 2017 an explorer on that expedition, Pen Hadrow, was part of another expedition that, hopefully given past experiences, tried to sale right up to the North pole. It was stopped almost 1,100 kilometres short of its objective by thick pack ice (www.arcticmission.com).

All these bung investigations come after nearly 30 years of

scientists forecasting the end of snow and ice, the melting of the Arctic and Antarctic ice caps and the disappearance of the Greenland ice sheets. As one example of this constant doom-saying in 2000 a Dr David Viner of the Climate Research Unit at the University of East Anglia in Britain told the Guardian newspaper that "children just aren't going to know what snow is". Heavy snowfalls would be a "rare and exciting" event.

Dr Viner, who went on to a senior role in Britain's climate change bureaucracy and then into consultancy work, was reacting to the unusually warm UK winter of 1999-2000. Unfortunately for him, in what turned out to be a public relations disaster for the global warming industry, most of the winters in the Northern hemisphere since then have been cold and snowy. (As noted above, those cold winters are really due to global warming, or so we are told.) Whenever British climate skeptics, in particular, have found themselves snowed in, they have recycled Dr Viner's forecast on their blogs. The mainstream media quickly forgot the interview and swallowed, almost without question, the environmental movement's relabelling of global warming (where was the warming?) as climate change.

Further along in time and closer to home, the so called Millennium drought in South Eastern Australia which lasted 12 years up to 2009 brought forth the usual crop of dodgy forecasts. In 2010, the South East Australian Climate Initiative, a joint venture between CSIRO and the Bureau of Meteorology produced a report 'Climate variability and change in south-eastern Australia'. The executive summary of this report from what are undoubtedly a host of distinguished scientists says this:

"In summary, to the extent that the current changes in temperature and rainfall are linked (at least in part) to climate change, it is possible that the climate in south-eastern Australia is shifting. This raises the possibility that the current dry conditions in south-eastern Australia may persist, and even possibly intensify. However, given that natural variability is also likely to be playing a role in the rainfall decline, it is also possible that there may be a return to somewhat wetter conditions in the short-term."

This rather cautious, qualified forecast was picked up by the media as an assertion that the drought would continue. This proved unfortunate as a few months after the printed report landed on the desks of the team that compiled it, the rains started. And it rained. The SEACI produced another report the next year which acknowledged the flooding, correctly attributing it to a combination of the La Niña and the Southern Annular Mode climate cycles (SAM), but maintained its cautious forecast. Most Australians will be aware of the la Nina and related el Nino climate cycles resulting from changes in ocean temperatures in the Pacific, but fewer will be aware of the SAM, which relates to the weather belt around the Antarctic. This belt can expand far to the North, and when it does rain comes with it. The two climate cycles can interact.

Unfortunately for SEACI's forecast, the Murray-Darling

system also flooded the next year, with the Bureau of Meteorology subsequently declaring the two years, from March 2010 (the release date of the report) as the wettest two years in South East Australia on record. Many commentators blamed the flooding on global warming.

As a journalist, I wrote a column well after the event pointing to the first report as a failure of forecasting. The report authors then sent me a note saying that they had also forecast flooding in the short term, which they had, very cautiously. However, rainfall patterns of subsequent years have shown no real signs of reverting to the conditions of the Millennium drought. Droughts have since come and gone – there is one now in parts of the South East - but they have been a major feature of Australian agriculture since settlement and most years since the millennial drought broke there have been floods somewhere in the Murray-Darling. As I write this in Sydney, rain is pounding on the roof, as it has done off and on for the past three days. This heavy rain is being driven by what is now called an east coast low, a low-pressure system that forms in the Tasman.

The breaking of the Millennial drought, incidentally, also marked a turning point in climate scepticism in Australia. When the drought was on, plenty of people were prepared to believe that climate was changing for the worse. When the drought broke they realised that the flooding in many areas was the same as the last big flooding episode in the 1970s. In Brisbane the flood peak was below that of the 1974 episode, but then substantial flood control works had been completed since then, including the Wivenhoe dam further up the Brisbane river. At the time of the floods, various commentators claimed that higher temperature of the coastal waters at the time meant more water vapour in the atmosphere which became rain. Unfortunately for that argument, as I noted, there is no evidence that the floods of 2010-11 were any worse that the 1974 floods. Elsewhere, people observed that the flood waters reached the same heights as previous episodes, which had been marked on natural features (I had conversations with people who have noted this), and began to wonder whether the endless pronouncements of doom and water scarcity on the ABC could be true.

As previously noted another serious drought now rules in South Eastern Australia, with the Bureau of Meteorology grandly noted that in 2018 "New South Wales had its sixth driest year on record, while the Murray-Darling Basin saw its seventh-driest year on record." Australia's rainfall in September, in that seventh driest year, was also the lowest on record. Without detracting from the seriousness of the drought for the farmers who have to deal with it, the records noted by BoM are hardly dramatic, especially considering that the bureau's records would not stretch back very far (the release declines to give the record cut off point).

While on the subject of droughts and floods, we should note the various forecasts of Australia's own climate prophet of doom, Tim Flannery. For Dr Flannery has a long and celebrated record of making gloomy forecasts that have proved to be wide of the mark. In 2007, when the Millennial drought was still in full swing, Dr Flannery warned that major cities like Brisbane would "never have dam filling rain again". In fact, he and other would be prophets, made so many alarmist statements during the drought while the state governments of New South Wales, Queensland, South Australia, and Victoria all panicked into building desalination plants at immense cost to the taxpayers. The plants in New South Wales and Queensland have never made it out of mothballs. The Victorian plant, which cost $4 billion, was used for the first time in early 2017, five years after it was completed, although there was no apparent need for water from it. As the plant requires 90 megawatts of power to operate at a time when global warming hysteria is forcing the state's brown coal plants to shut, the installation can be described as worse than useless. The plant in South Australian, which cost $1.8 billion to build, is run at minimum capacity to avoid maintenance issues, with SA Water conceding that it is not needed. Water bills in the state have doubled.

In contrast, the state of Western Australia has built two desalination plants which are being put to good use, and is planning a third. But the rainfall story in that state is different with noticeable reduction in rainfall in the past three decades. This reduction in rainfall has, of course, been attributed to global warming, although only WA seems to be affected. Another major suspect is a long-term climate cycle, which is a form of climate change but is not driven by changes in carbon dioxide content. West Australians should invite prominent climate change activists to make pronouncements about how the state's climate is becoming drier and then watch the rains start again.

As we can see, and as any layman would suspect, forecasting long-term rainfall patterns on climate models that are still largely incomplete is so much wishful thinking. Stewart Franks, a Professor of Environmental Engineering at the University of Tasmania who specialises in studying just how climate variations can affect the hydrology (floods, droughts, soil moisture) of the landscape, has pointed out forcefully in interviews that climate models cannot yet simulate the El Niño and La Niña cycle in anything like a realistic manner, or the clustering of those events, so how can they say anything about future trends in floods or droughts?

Another point that can be made about all of this is that the claim about "settled, certain" science, was only meant to refer to temperature forecasts. Long term forecasts of rainfall patterns are a different beast entirely. Activists are assuming that higher temperatures must mean reduced rainfall. But anyone who has been to a tropical country know that it is not as simple as that.

4

The End of Days Very Soon

"If the apocalypse comes, beep me!"
Buffy the Vampire Slayer (said at a time when pagers were cool).

Fears about climate change have provoked all sorts of wild and wonderful forecasts that could well have fitted into an episode of the now venerable Buffy series, in which the apocalypse was always about to happen unless Buffy and her friends prevented it.

Climate refugees

Environmentalists are not happy unless they are forecasting that storms, floods or just increasing temperatures and rising sea levels will drive millions from their homes. One of the chief doomsayers in this area is Norman Myers, a professor at Oxford University and an environmentalist who is deeply concerned with climate refugees. In 2005, he forecast that there would be 50 million such refugees by 2010. That number would rise to 200 million when climate change really started to take hold.

The UN took up Professor Myers' forecasts with enthusiasm. In the same year the United Nations Environment Program (UNEP) issued an identical forecast. In 2008, Srgjan Kerim, president of the UN General Assembly at the time, declared that there would be between 50 million and 200 million environmental migrants by 2010.

Sadly for those distinguished scholars and officials, few other scholars specialising in geography or migration would have anything to do with Professor Myers forecasts. We are now well past 2010 of course, with no such dislocation evident. Various media articles have pointed out that in many of the areas where increasing sea levels are supposed to be driving people from their homes, such as the Bahamas and the Solomon Island, population has been increasing, not decreasing.

This is also the case in the island of Tuvalu, an independent nation about half way between Australia and Hawaii which, we have frequently been told, is about to sink under the waves. The island is certainly low lying – its highest point is less than five meters above sea level – but there is no sign of a mass migration off the island, despite claims by the island's government that the situation is desperate. A 2002 census put the population at 9,561, and a 2015 estimate was 10,869. That is an increase of more than 10 per cent in 13 years.

Activists have tried to link climate change to the mass refugee flows that occurred into Europe in the past few years, mainly from Syria due to the civil war there but also from countries of the Middle Eastern and Northern Africa. A drought in Syria probably didn't help, but there have been droughts in the area before without the country falling apart. Political instability trumps all other causes in refugee issues. At the time of writing Syria's seemingly intractable civil war still rages.

This history of failed forecasting has not stopped activists from making ever more dramatic claims. In June 2017, a group known as the Breakthrough Centre for Climate Restoration released a report entitled Disaster Alley which claims that climate change could be a threat to intelligent life on earth. The report sets out a number of dire scenarios including the possibility of sea level increases of tens of meters displacing millions from coastal cities. However, these scenarios are expected to play out over decades, rather than just a few years as with the earlier forecasts.

Not to be outdone by a mere group of activists, also in June 2017 the venerable Cornell University in the US released a report forecasting that 1.4 billion people or well over 10 per cent of earth's population would become climate refugees by 2060, due to rising sea levels. The figure would top two billion by 2100. This report is, in turn, based on another estimate of likely sea level increases (see later in the chapter).

In a few years, when nothing much has happened both Cornell and the Centre can release more reports saying the same thing but with a later catastrophe start date.

Sea levels

This is the oddest of all the scare stories, in that the global changes in sea levels have been tracked by satellite for the past twenty years or so and are well known, but still activists, some scientists, and even The Pope persist in ignoring the actual measurements, which show little sign of any unusual acceleration and instead react to dire forecasts.

The worst-case scenario set out in the 2014 IPCC report was for an increase of 74 centimeters by 2100, but that was not considered alarming enough for the many activists who infest this area. In May 2017, the US National Oceanic and Atmospheric Administration produced updated modeling taking into account the work on the melting of ice sheets and so on, which came up with a sea increase of more than two meters by 2100. Much better! Newspapers such as the Melbourne Age then ran stories saying that a large part of Melbourne (undoubtedly a flat place) would be inundated by such an increase. Something must be done about emissions!

As I write this I am sitting in the top story of a two story house in Melbourne well away from the nearest sea water in Port Phillip Bay (other parts of this book were written in Sydney), but I am assured that if sea levels increase by two meters and nothing is done about it, the bottom of the house will be underwater. Very well, but before we all start selling and moving to higher ground, consider the time period. The year 2100, or more than 80 years from now, is about as distant in time as the lead up to the second world war. Babies born now will be elderly by the time these dire predictions come to past (assuming that aging has not been conquered by then) and society will be quite different. How will it be different? I have no idea, but the house I am in will have long since been knocked down and replaced, as will plenty of other dwellings, including many on the foreshore.

Now consider that sea levels are tracked by satellite – several satellites using lasers - and the graphed results can be seen by all at www.columbia.edu/~mhs119/SeaLevel/. According to those satellite readings sea levels have increased by maybe 80 millimeters (eight centimeters, or about half the length of an ordinary paperback book) since 1993, when measurements started, an average increase of 3.1 mm a year. There is no evident acceleration in the satellite readings, at least to the untrained eye, although scientists have been busy "re-analysing" the date to find an acceleration.

If you look further down the same page produced by Columbia University, you will also see that scientists have matched the satellite readings to tide gauge measurements (from instruments attached to piers and the like measuring increases year on year). In the thirty years up to 1930, according to those instruments, the sea level increase was 0.6 mm a year, with the satellite data showing an increase of 3.1 mm a year, or so that story goes. There you are, there must have been an acceleration, and never mind that it's not evident in the satellite results.

Skeptics will mutter about the problems of matching measurements from earth based instruments (the tidal gauges) to satellite readings, and about whether the change is really due to industrial activities or due to a climate cycle. Perhaps. The reality of sea level increases is that we can see for ourselves what's happening year on year right now and do the math.

If nothing else happens the present annual increase in sea levels, to take simply the satellite readings and ignore attempts to match them to other measurement systems, will result in a piddling additional 25 centimeters by 2100 (less than a foot in the old Imperial measures) and anyone living in whatever buildings occupies the site on which this is being written will not have to wade around their living room. If sea level changes accelerate, it will take decades to build up into a significant increase. That means there will be time to adjust foreshore building codes to suit, or perhaps even build barriers if that is necessary. Foreshore structures can then evolve naturally as they are knocked down and replaced to accommodate any increase. Bear in mind that current designs already allow for a certain increase. Will it be worse than that? It just does not matter.

Then there is the question of whether the increase we can see now is due to any human influence or is simply part of a cycle in changes in sea level. See what the IPCC now says about rising sea levels in a later chapter.

Diseases

When the Ebola virus caused many thousands of deaths in West Africa – notably in Guinea, Liberia and Sierra Leone, in a bad two and half year outbreak up to the beginning of 2016, experts pointed the finger at climate change. In 2014, just as the Ebola outbreak was kicking into gear, a publication by Columbia University in the US called *State of the Planet* said:

"Some scientists think that climate change, with its increase in sudden and extreme weather events, plays a role in Ebola outbreaks: dry seasons followed by heavy rainfalls that produce an abundance of fruit have coincided with outbreaks. When fruit is plentiful, bats (the suspected carriers of the recent Ebola outbreak) and apes may gather together to eat, providing opportunities for the disease to jump between species. Humans can contract the disease by eating or handling an infected animal."

Without detracting from the expertise of the scientists who have put this report together you will note that they are pointing to a sequence of events – a dry season followed by heavy rains – and that sequence is somehow promoted by climate change. Leaving aside the issue of whether this climate sequence is more likely due to whatever climate change is supposed to mean, a point carefully left out of the report is that any resulting increase in the number of disease would have to outpace improvements in medical services and technology to make any difference to death rates.

In Western countries the figures speak for themselves, including the General Record of Incidence of Mortality (GRIM) workbooks, maintained by the Australian Institute of Health and Welfare, a government agency. These show that in 1907 the death rate due to infectious diseases of all kinds in Australia was 283 per 100,000. In 2014, the latest year for which figures are available, the death rate was 12 per 100,000.

West Africa obviously has nothing like the medical services or economic development of Australia, not to mention up to date statistics, but a World Health Organisation statistical profile for Guinea for 2014, before the Ebola outbreak, shows that many thousands of Guineans were dying from HIV (which is still rife in the area), tuberculosis and malaria, and many thousands more due to "other" infectious diseases. Thanks to the excellent work of WHO and other bodies, the death rates from those disease have been declining for years. Just how much the Ebola outbreak would have affected those figures is very hard to say but as far as anyone knows the death rate was in the thousands, not tens of thousands. Now that the outbreak has passed, the basic work or reducing death rates from all other causes can continue. Climate change does not enter into it.

A global pandemic is possible, just as anything is possible, but it is difficult to see how any new infection would get past Western disease control systems to the extent of making any significant difference to death rates. Even AIDS/HIV which did gain a hold in Western countries and took some time to control probably made little difference to the overall decline in deaths due to infectious disease.

A better example of just what happens when a disease kicks off is the 2009 flu pandemic or swine flu outbreak. This was undoubtedly a nasty disease for those who contracted it, but the terror it invoked in advanced countries seemed to be out of all proportion to the danger it presented. The health authorities took strong action, which included preventing children switching schools from attending their new school for some days, until they were declared disease free.

Great Barrier Reef

The Great Barrier Reef has been in trouble ever since I can remember. In the 1960s the threat was from the Crown of Thorns starfish, now its bleaching due to climate events. Every now and then there are news items declaring that the damage is worse than previously expected. On August 3, 1971, the Sydney Morning Herald declared that the reef would be dead in six months. In 2012 reef "experts" claimed that the reef had lost 50 per cent of its coral over the preceding few decades (*The 27-year decline of coral cover on the Great Barrier Reef and its causes*, the Proceedings of the National Academy of Sciences). The paper says that if nothing is done, just 5-10 per cent of the coral will be left by 2022. In 2017, or five years later, the claim was the same but over a much shorter period (*Half the Great Barrier Reef may have died in last two years*, News.com May 23, 2017). The reports note that estimating the extent of the damage is difficult, but then according to earlier forecasts there isn't supposed to be much left of the reef by this stage. So, the reef isn't dead yet?

A check of visitor numbers for the marine park which includes the reef, compiled by the Great Barrier Reef Marine Park Authority, shows that visitors are flocking to the reef in ever greater numbers. In 2010, the authority counted 1.6 million visitor days, but by 2016 it clocked a record 2.4 million visitor days, with tourists gazing at the attraction that scientists have long forecast shouldn't be there by now. That major change in visitor numbers probably has little to do with the health of the reef. A more likely explanation is the fall in the value of the Australian dollar, and an undoubted sharp increase in international travel by the Chinese, who have apparently failed to realise the reef isn't supposed to be there.

There are indications that the constant pronouncements of the death and/or immense damage done to the reef by global warming/climate change is having some effect on visitor numbers if nothing else. In early 2018 a leading light in the Queensland tourism industry, Col McKenzie, blamed a downturn in visitors to the reef on "misleading" statements by scientists about damage to the reef. To sort through claim and counter claim in this area would be dull indeed, but the GBR is a vast natural feature which survived the last ice age and previous intergalacial (the period between ice ages), which is widely accepted to be hotter than the present intergalacial, and its still there. The changes of the last 30 years or so would seem to be small beer in comparison to what the reef has already survived. For some further discussion on the science behind the continual pronouncements of doom of the roof, see the chapter on experts.

Food security

Despite Australia's role as a major food producer and exporter, there are those who insist that really it is on the verge of starvation. When food prices spiked in 2011 it was possible to find senior academics talking about the need for food security for major Australian cities. Say, what?

Unfortunately for this particularly nutty brand of doom saying, including forecasts about how climate change will knock agricultural production for a loop, food prices have refused to follow the script. A UN body, the Food and Agricultural Organisation, compiles food indexes and the overall food index shows a price spike in 2011. That was due to a number of reasons way too boring to discuss here, but include the rise in food consumption in China due to that country's improving economy. Fewer people in poverty means that more food is eaten.

Since then, however, prices have been mostly falling (they have picked up in recent months). The FAO adjusts the index for inflation which shows that in real terms food prices are back to about where they were in the mid-1970s. That is not what doom-sayers want to hear of course, but a closer look at the graphs show that after decades of forecasts of famine, kicking off with Professor Paul Ehrlich's 1968 book *The Population Bomb*, the behavior of food prices might have changed. Up to the mid-1980s farmers had to increase productivity just to keep ahead of prices that were declining in real terms, but from the mid-80s to the big run between 2005 and 2011 they were basically static. Despite the recent decline in prices, the index remains above those previously static prices. In other words, food is generally more expensive than it was before the crisis.

What does that mean? I have no idea and most of the academic discussion of this issue dates from 2011 when food prices were spiking. At that time everyone was predicting prices would go higher. But higher prices generally mean more food production. As the European Union's experience with subsidies to farmers has shown, show farmers a dollar and they will produce any amount of food we want, and never mind climate change.

Bung reports

There have been lots of these but perhaps the most interesting was one released by the Pentagon in October 2003 entitled *An Abrupt Climate Change Scenario and Its Implications for United States National Security.*

This report sets out a scenario where climate flips, so that large areas become much colder and others much warmer due to disruption in the usual circulations patterns in the ocean and atmosphere caused by warming. As the report was released more than a decade ago without the disaster it foresaw happening we can now safely ignore it. One author has since commented that he was astonished at the attention it generated, saying, in effect, that it was 'just a scenario that we put forward'.

Deadlines and tipping points

This is worth a whole section to itself as scientists and activists have been freely forecasting a deadline for action or a tipping point for the earth's climate is just five or ten years away for decades. This is just a small selection of the warnings.

In the *Miami Herald*, July 5 1989: *"A senior U.N. environmental official says entire nations could be wiped off the face of the Earth by rising sea levels if the global warming trend is not reversed by the year 2000. Coastal flooding and crop failures would create an exodus of "eco-refugees," threatening political chaos, said Noel Brown, director of the New York office of the United Nations U.N. Environment Program, or UNEP.*

In *The Guardian* on May 5, 2007. *"Governments are running out of time to address climate change and to avoid the worst effects of rising temperatures, an influential UN panel warned yesterday.*

"Greater energy efficiency, renewable electricity sources and new technology to dump carbon dioxide underground can all help to reduce greenhouse gas emissions, the experts said. But there could be as little as eight years left to avoid a dangerous global average rise of 2C or more."

In the *New York Times*, June 1, 2007: *A stern warning that global warming is nearing an irreversible tipping point was issued today by the climate scientist who the Bush administration has tried to muzzle.*

James Hansen, director of NASA's Goddard Institute for Space Studies in New York, today published a study showing that greenhouse gases emitted by human activities have brought the Earth's climate close to critical tipping points, with potentially dangerous consequences for the planet."

The Guardian, January 18, 2009: *"Barack Obama has only four years to save the world. That is the stark assessment of Nasa scientist and leading climate expert Jim Hansen who last week warned only urgent action by the new president could halt the devastating climate change that now threatens Earth. Crucially, that action will have to be taken within Obama's first administration," he added.*

In the ABC, October 19, 2009 – *"A new report has given world leaders a deadline of 2014 to embrace a low-carbon economy or see the planet hit a "point of no return".*

The economic modelling, commissioned by WWF Australia, has found that an emissions trading scheme is not enough to drive the change needed to sufficiently cut global emissions."

The UK *Telegraph*, November 15, 2009: *"Pollution needs to be brought under control within ten years to stop runaway climate change, according to the latest Met Office predictions.*

In the first study of its kind, climate scientists looked at how much pollution the world could afford to produce between now and the end of the century in order to keep temperature rises within a "safe limit".

The *HeraldSun*, December 4, 2009: *"The planet has just five years to avoid disastrous global warming, says the Federal Government's chief scientist.*

"Prof Penny Sackett yesterday urged all Australians to reduce their carbon footprint."

An honourable mention for persistence must go to those who signed the article *World Scientists warning to humanity: a second notice* which appeared in the journal *Bioscience* in November, 2017. This appeared 25 years after a first warning with much the same content and with formidable scientific support. Like the first warning, the Bioscience article has gained more than 1,000 signatures. The warning is the usual stuff about the earth's climate and ecosystems being in crisis. However, unlike the first warning no-one seems to have paid much heed to it, except to comment that over the top warnings of climate doom may do more harm than good.

Just to show that warnings about climate have not fallen out of fashion, the IPCC released a special report on global warming of 1.5 degrees centigrade in October 2018 which claimed, you guessed it, that the climate story is worse than was previously though and that considerably more action was required to avoid tipping points and so on. In other words, its findings are just like all the other reports issued in the past thirty years.

There have been several stories over the years forecasting a tipping point in climate alarmism, where all these doom-sayers realise that their endless warnings are turning people off the climate change story. Sadly, these have also proved wrong.

5

Experts are not everything

In my time I have been threatened by experts. And I don't rate you very highly at all.

- Doctor Who (Tom Baker years).

In 1999, as editor of a specialist financial magazine I found myself in a round table with a senior member of a major accounting firm (I won't give names to protect the guilty) and various senior financial officers of large companies on the issue of the Millennium bug. This was the fear that many computers would stop working when the new century came in on January 1, 2000, as they had not been built to take into account the change to all four digits in the year's date, instead of just the last two. One of the senior financial officers, asked me if I thought the disruption on January 1, in just a few months, would be serious. I responded that I did not think that it would. I was immediately contradicted by the senior member, whose firm had been making good money consulting on changes in systems to accommodate the change. He insisted that the disruption would be substantial.

On New Year's Eve I saw queues of cars at the service stations as people made sure they had enough petrol to ride out any disruption and I found a horde of long lasting milk stacked under the family kitchen sink. In the end, as is well known, nothing whatever happened. Even the venerable 286 PCs of the time were completely unaffected. One common excuse for this failure of prediction is that a lot of money was spent fixing the problem before it occurred. But as noted in the book *Future Babble Why Expert Predictions Fail and Why We Believe Them Anyway* by Canadian journalist Dan Gardner, there was no apparent correlation between problems that didn't occur, and money spent on fixes beforehand. British Telecom spent vast sums on Y2K fixes while the equivalent organisation in South Korea, using much the same equipment on the same scale, spent nothing and got exactly the same level of disruption.

As Gardner notes, there were people who investigated the issue, came to the correct conclusion that there was no real problem and tried to tell the media, for very little result. For every expert who had reached the correct conclusion there were many more experts with impressive credentials prepared to confidently forecast, in front of a camera, that a disaster was about to occur. To these can be added the many activists who don't know much about the issue, but are firmly convinced that disaster is about to happen. In any case, forecasts of disasters make for better stories.

All this meant that even senior information technology and finance executives who, in the 1990s, might have suspected that the whole Millennium bug story was nonsense would have been reluctant to stand up to a board of directors who had just seen a news item about a looming disaster and wanted to ensure that their company did not get caught. In any case there was nothing wrong with updating the computer system and if anything did happen, the executive was covered. (A similar mechanism is behind the occasional outbreak of new disease stories, incidentally. The worst possible scenario becomes the only possible outcome and the authorities then have to be seen to be taking it seriously, if only to avoid mass panic.)

After the Millennium bug deadline passed with nothing at all happening, the reaction of the "experts" to being proved completely wrong about that and other supposed crises, as Gardner noted and I have also observed, was to brush off the failure with a glib remark and move on. Another response was to simply claim that they were right. One poster in an online forum, when I mentioned the Millennium Bug some years after the event, took another approach by indignantly denying that there had been much fuss in the first place. Maybe there had been a few stories and warnings, but not the endless flood that those who lived through the event remember. That meant it could hardly be a failure of prediction.

Perhaps in a few years those now shouting about how global warming/climate change will get us all and that the end is nigh, will claim that there was not much fuss, maybe a few stories here and there, but nothing really worth mentioning. Right!

Despite "experts" being completely wrong about Y2K and many other subjects, one often repeated claim about the global warming/climate change story is that so many scientists agree that it is right, that there must be something to it. The argument about expert authority is often used to shut down debate. These people with impressive credentials say something is right, then it must be right, so can all ignorant sceptics go away?

First off it's not true that all scientists vouch for climate change. There is an honourable opposition led Professor Richard Lindzen a professor emeritus (meaning retired) at the Massachusetts Institute of Technology, where he was formerly professor of meteorology. But it is true that the orthodoxy of the moment is that increasing carbon dioxide from industrial emissions is supposed to be causing increases in the temperature, or heat, of the earth, and there is little apparent dissent.

One of the reasons that there is so little apparent dissent, however, aside from honourable exceptions such as Professor Lindzen and Will Happer, a physicist at Princeton University in the US (who takes the view that the warming effect has been greatly overstated), is that working scientists in the field don't dare disagree. Note the case of Bob Carter, a marine geologist and formerly a professor at James Cook University in

Queensland. Professor Carter, who died in early 2016, was an outspoken critic of climate change orthodoxy and one of the leading Australian sceptics. Although he had retired he still held the position of adjunct professor (meaning not paid) with the university's School of Earth and Environmental Studies. He also gained some consulting work paid through a research grant to a colleague and supervised a student. It was too much for JCU. In 2013 the arrangement was discontinued. The university was careful to say that they did not disagree with Professor Carter's views. The problem was that those views did not fit with those of the university and so were difficult to defend.

The James Cook University has also parted ways with one of its academics, a Professor Peter Ridd, who criticised the quality of Great Barrier Reef research on Sky News in August 2017. JCU thought the comments concerning the institutions with which it is linked, the Australian Institute of Marine Science and the ARC Centre of Excellence for Coral Reef Studies, worthy of issuing Professor Ridd with a formal censure and gag order which the academic has refused to accept. At the time of writing the dispute was heading for the Federal court and Professor Ridd had left the university.

Without commenting further on the JCU case, expressing doubts about global warming is never a good idea. The best known, and earliest, example of discrimination against sceptics concerns Hendrik Tennekes, a former director of research at the Royal Dutch Meteorological Institute (the Dutch Met office). He was forced out of his job at the Dutch Institute in the 1990s for daring to speak out against climate change dogma.

Closer in time and distance, Garth Paltridge, emeritus professor of the University of Tasmania and a former director of the Antarctic Co-operative Research Centre, says that when he was setting up the centre in the early 1990s he made the mistake of saying to the media there was still a lot of doubt about the science of global warming. In an article in the *Australian Financial Review* (February 22, 2006) he says he was quickly told by executives at the "highest levels" of the CSIRO — a partner in the centre — that he should keep such doubts to himself or see the organisation withdraw from the centre. At the time the CSIRO was trying to get tens of millions of dollars of funding from the newly formed Australian Greenhouse Office (which has since been abolished).

Ordinary scientists have also complained that the moment they express any doubt about the theory of climate change then colleagues will not work with them on papers, or respond to queries. For climate change stirs passions unlike almost any other scientific issue, both among scientists and the mass of activists who don't know anything much at all about the science, but know that it must be right because it fits with their beliefs about the over-use of resources and greedy capitalism. I have experienced this directly myself in news rooms, which are filled to be brim with activists.

One other area that seems to cause similar patterns of emotional response and discrimination is that of smoking. There is anecdotal evidence that any scientist who suggests that secondary smoke inhalation may not be instantly deadly (I exaggerate here) will find themselves ostracised by their peers. To avoid having the anti-smoking lobby in my face screaming that they don't get emotional over this issue, I declare that I have never been a smoker, care nothing about the debate one way or another and won't respond to any attempt to draw me into it.

Another illustration of the passion and prejudice against sceptics that the climate debate seems to invoke is that of *New York Times* (NYT) columnist Bret Stephens. A former Wall Street Journal writer he wrote a column for the NYT which the skeptics thought was mild. As far as Stephens is concerned, climate has been changed by human influence, but citizens had a right to be skeptical about what he describes as "overweening scientism". This is hardly remarkable stuff but at the time of writing a petition on change.org (a web site where activists can organise petitions) calling for *The Times* to sack Stephens had gained more than 40,000 signatures. Usual suspects among the scientists such as German professor Stefan Rahmstorf have written to the *Times* to complain.

Those who write newspaper columns promoting extreme, alarmist views of climate change do not stir up this sort of outrage, but activists will tell you that exactly the opposite is the case. They claim that those fearlessly investigating climate change have been hampered in their work and prevented from speaking by shadowy forces, funded by sinister big-energy companies. (While we are on this subject does anyone know how I can get funding from big-energy?)

Perhaps the best way of rebutting this is to point out that the climate change guys win awards and have documentaries made about them – documentaries which show the sceptics as the bad guys funded by those same nasty fossil fuel interests. Then there are whole films showing that climate change will ruin the world, and that the deniers are mentally troubled. These media efforts include *An Inconvenient Truth* (2006) about the efforts of former US president Al Gore to educate the masses about global warming; the UK produced *The Age of Stupid* (2009) which depicts a world of 2055 ruined by climate change; *Earth 2100* (2010), a two-hour television documentary exploring the worst case scenario of what will happen if we don't take action against climate change; the documentary films *Chasing Ice* (2012), *Thin Ice* (2013) and *The Expedition to the End of the World* (2013) which is about the Greenland ice sheet. There is also *Snowpiercer* (2013), a Czech-Korean film featuring mainly English-language actors about a climate apocalypse. That last example is perhaps not so relevant as climate apocalypse films were being produced long before the global warming story came along, and will be produced long after it is gone. More to the point is the short, satirical film *Climate Change Denial Disorder* (2015).

In 2017 Former US Vice-President Al Gore returned to the fray with *An Inconvenient Sequel: Truth to Power* but there were indications that the mainstream media and the public had become tired of apocalyptic climate warnings. The film's reception and box office takings have to date been far less than the original.

Against all of that plus various film references where those working against the climate change story are the bad guys, to be beaten up by the likes of Supergirl, skeptics can only point to the 2007 British documentary *The Great Global Warming Swindle* which was bitterly attacked the moment it was aired.

Those pushing the climate change story are the persecutors not the persecuted.

All of that said, with President Trump in power global warming proponents have been getting a taste of their own medicine, and they are not happy. Even before President Trump's inauguration in early 2017, government scientists who come to prominence during the Obama administration, were busy declaring that science would be repressed, and that key climate data collections would be destroyed or altered. At the time of writing the web site of the main battleground agency, the Environmental Protection Agency, still features plenty of climate propaganda, and the concerns about databases have been forgotten (the Trump administration has no reason to mess with them). President Trump has, however, cut the agency's budget by one third and initially appointed as its head Scott Pruitt, whose previous career as attorney general for the state of Oklahoma mainly involved attacking the EPA in law suits.

Mr Pruitt has since left the agency, after a series of questionable decisions, but journalists who have spent most of their careers denouncing climate sceptics are now attending press conferences run by the very sceptics they have been denouncing, to the evident horror of those journalists.

Another point to make about the authority of experts is best illustrated by a conversation I had with a former colleague and a fervent believer in climate change. This colleague also happened to mention to me that she had doubts about vaccination. She had been listening to fringe groups campaigning against the medical procedure. I told her that I was horrified. Vaccination, to talk of the medical procedure as one process (of course, there are many different forms of vaccination against different diseases) has a solid, well established track record of achievement. Before the advent of systematic inoculation against diseases, for example, children use to die like flies.

In other words, my colleague was enthusiastically embracing what experts have to say in one area, global warming, where the theory has no forecasting track record and no achievements, but expressing doubt about expert opinion in another area where, contrary voices are almost non-existent, and there is a solid track record of achievement. Why? In both cases my colleague was listening to horror stories – taking the counsel of her fears. Never mind the science, which she knew little about, just listen to the horror stories.

That is an exact reversal of what should be happening. In astronomy, for example, you would pay attention to what astronomers say when they forecast the time of an eclipse of the moon and sun. These matters are well understood and have been forecast accurately now for centuries. Prognosis of a heart condition is still partially an art rather than a science, but when a properly qualified heart specialist says you have a problem you will listen because he or she is drawing on a substantial body of knowledge, where forecasts have been proven correct and risks have been properly quantified through extensive surveys and trials of drugs.

As discussed in other chapters all climate scientists can point to is a number of dodgy forecasts, including a string of long-gone tipping points, but they are still claiming the same status for their forecasts as predictions about heart disease, or eclipse times.

Another often used excuse for why anyone should pay attention to endless forecasts of climate doom, is that the papers containing the warnings have been peer reviewed. The answer to that is so what? Peer review is a process used by scientific journals for weeding out papers that do not add to the debate to the scientific field in question. The editors of the journal send papers they are considering for publication to two scientists in the field who then make comments on it. The paper's authors are not told who the reviewers are. Depending on the comments, the paper is accepted, revisions are requested before it is considered again, or the paper is rejected. No one has ever previously suggested that merely because the paper has been accepted after peer review that it is right, merely that it is worthy of consideration. As is well known a lot of flawed research has been let through.

This is not to criticise peer review. The process has served the scientific world well enough and no-one has been able to think of a better system. The problem is when this review process is used to justify a forecast. What it means is that the forecast fits with the available scientific theory. How do we know if the theory is correct? When the forecast turns out to be correct.

In the mean time scientists can easily be collectively wrong because the theory was wrong, with the classic example being that of stomach ulcers.

In 1982, two Australian scientists, Dr Robin Warren and Dr Barry J. Marshall, identified the bacterium *Helicobacter pylori* as one of the major causes of stomach ulcers, not stress or spicy foods as had previously been assumed. Up until that time and for years afterwards, experts would have told anyone afflicted with a stomach ulcer that it was about spicy foods and stress, and they would have been wrong. Dr Marshall has since declared that the bacterium hypothesis was ridiculed, but those with oversight of the field have bitten back saying that the hypothesis was the subject of lively discussion from the beginning. The problem was that establishing that a specific microbe as the cause of the disease was not simple. After some years and an incident in which Dr Marshall famously swallowed a petri dish of the bacterium and developed an ulcer, the criteria was met. Then it was accepted.

Whatever anyone may make of this issue, readers will note that up until the 1980s, doctors were wrong about the cause of stomach ulcers because the theory was wrong. The issue was corrected but not after some years and when certain rigorous criteria was met. However, everyone was clear about those criteria and knew when the standard of proof had been met. There are no such standards in climate science. In that branch of science, it seems, a result is right because scientists say it is right. Why? Because they are experts.

6

Meetings and yet more meetings

"I think there needs to be a meeting to set an agenda
for more meetings about meetings."
- Jonah Goldberg (American columnist and author)

Who is tired of hearing about meetings which aim to produce
an international agreement on curbing emissions? Every year or
so, tens of thousands of delegates will congregate in an exotic city,
having expended lots of emission to get there on jets, to try to
hammer out some form of accord on controlling emissions. Apart
from the government delegations, which collectively number in
the thousands, there are typically many more thousand from
non-government organisations of all types and sizes which have
vowed to fight against climate change.

David Henderson, a former head of the Economic and
Statistics Department at the OECD who has been very critical
of the IPCC's attempts at economic analysis, once estimated that
23,000 representatives from NGOs turned up at the Copenhagen
meeting in late 2009. The vast bulk of these would have been
climate change activists. There have since been meetings at

Cancún, Mexico (2010); Durban, South Africa (2011); Doha, Qatar (2012); Warsaw, Poland (2013); Lima, Peru (2014) and Bonn, Germany in October 2015 which produced the draft agreement for the Paris meeting in late 2015. Press reports put the number attending Paris at 50,000, or more than double the number attending Copenhagen. Again, almost all would have been climate activists. Sceptics know they would have been lone, derided voices and would have had to pay their own way. There is no funding for sceptics.

The meetings keep on coming. There has since been another gathering of activists and negotiators at Marrakesh, Morocco, Suva in Fiji and Katowice in Poland. Listing the 13 major meetings before Copenhagen, starting with Berlin in 1995 and including the famous Kyoto meeting in 1997, would bore the reader.

What has all those additional jet emissions and hot air from endless discussion achieved? Very little and perhaps nothing at all. However, the real irony is that in the past couple of years emissions have stabilised for reasons that have nothing to do with the efforts to reach an international agreement. We will return to that point. For now, we will look at the only two agreements produced by this endless string of meetings that are worth anything at all – the Kyoto Protocol, which resulted from the meeting in Kyoto, Japan, and the Paris agreement emerging from the Paris meeting in 2015.

The Kyoto Protocol was never intended to achieve very

much, as those involved with it will freely admit. Hammered out during late night sessions, the agreement only covered developed countries excluding the US which never ratified it. Crucially it also did not cover the big emitters of India and China. Even those who signed the treaty did not have to do anything about it for many years as it was not ratified. Once treaties are signed by a government representative, say a foreign minister, the governments involved still have to formally assent to the treaty – to ratify it. The process for this varies with the government. The Kyoto Protocol, as the treaty itself specified, did not come into effect until 55 of the signing countries and countries representing 55 per cent of the emissions covered by it. The conditions for this were not finally met until March 2005 when Russia ratified, reportedly in exchange for the admission to the World Trade Organisation as a developing country, and despite strenuous objects from the country's scientists (Russian scientists have been notable hold outs to the prevailing endorsement of climate change theory). Australia did not ratify until the election of the Labor government of Kevin Rudd in 2007.

A lot could be said about how those who signed the protocol met its requirements without much pain. Britain's electricity industry at the time was switching from coal to gas from the North Sea (the country's electricity grid now uses very little coal). Germany and the former Soviet bloc states set their base lines before their old emissions-intensive Soviet style industries were swept away when the Berlin Wall fell. When Australia did finally ratify, it was helped in meeting its obligations by an allowance for land clearance and forestry. However, further discussion of the Protocol is not worth our trouble as it expired, as originally intended, in 2012. Exhaustive negotiations patched together an extension of the treaty known as the Doha amendment or second extension period, but this has yet to be ratified. A statement issued by the UN in July 2016, four years after the deal was hammered out, states that only 66 of the 144 countries required to ratify before the agreement can take effect had done so. Nothing much has happened since then.

Instead, the countries involved may default to the Paris treaty or agreement which has met ratification requirements and will come into effect in 2020, or well after the climate is supposed to be have tipped us over into ruin. This was hailed as the first legally binding climate treaty covering the bulk of world emissions, which is true up to a point. For the agreement legally binds the countries to produce an emissions target, the form and nature of which is up to the individual countries. The Paris agreement does not say anything about what the target should contain or require the countries involved to meet it. They simply have to produce a plan. If the target is thought inadequate by others or it is not met, then the country can be named and shamed, but that's about it.

Since the Paris meeting, even this limited agreement has started to unravel. One of the top emitters, the US, has withdrawn thanks again to President Trump. He was elected with the avowed intention of repudiating the Paris deal, and has since done so, much to the disgust of the climate establishment. Brazil (seventh largest emitter) may also withdraw thanks to new right-wing president Jair Bolsonaro. Other top emitters, China, India and Russia have carefully arranged their national pledges so that they do not have to do anything at all. India's target is particularly outrageous, as it includes a pledge that 40 per cent of installed electricity generating capacity would be non-fossil fuel by 2030 – a pledge which sounds impressive but is essentially meaningless as it is about installed capacity, not power delivered. According to the Indian Central Electricity Authority the 5 per cent of solar and wind energy now delivered to the national grid comes from 17.5 per cent of installed capacity, with hydro power (which counts as non-fossil) making up another 14 per cent, nuclear another 2 per cent and biomass (burning wood amongst other things) 2.6 per cent. That adds up to slightly more than 36 per cent. Finding an additional four percentage points in a decade or so will not be a burden, particularly as the country is already planning a major expansion of its hydro facilities.

As for the other major emitters, to expect countries like the Ukraine, Turkey, Saudi Arabia, Argentina and South Africa to do much in this area when they all have substantial, intractable domestic problems is basically an exercise in wishful thinking. A project known as Climate Action Tracker run by a consortium of climate research organisations in the US, looked at the progress the bulk of the major emitters in meeting their Paris obligations and, in a report issued late last year, concluded that few countries were doing anything.

In any case, the developing economies have always shown considerably more interest in a mooted $100 billion a year fund for helping them adjust to a carbon free future, than in setting tough climate targets. This fund was another feel-good activist initiative first suggested at the 2009 Copenhagen conference under the title the Green Climate Fund. The concept gained traction as it sounded like an excellent idea to everyone who did not have to contribute to it. Along the way it gained the aspirational goal, and it was just an aspiration, of $100 billion a year by 2020.

As of early 2019 the Green Climate Fund web site says that it has $US4.5 billion in committed funds, with those funds in various stages of being funnelled into feel-good projects. That is all very nice and this may even help poverty in developing countries, but that does not equal $100 billion and the developing countries have noticed.

This means that the fund, which started as a feel good initiative, has developed into a major bone of contention and a potential excuse for developing countries not to bother with their emissions targets. As far as those countries are concerned, as has been frequently reported, the Western countries caused this problems in the first place with their active industrial sectors, so they should bear the brunt of fixing it and pay developing countries to make the effort. If there is no money, why should developing countries bother?

While all this has been going on total emissions have not been increasing very much at all. A report by the Global Carbon Project, which tracks global emissions, states that global emissions from the combustion of fossil fuels and industry were expected to grow by more that 2 per cent during 2018, but that there had been no growth in the previous three years.

Various trends noted in the GCP report is the end of the construction boom in China, reduced smelting due to weaker global demand for steel, and the US electricity industry switching from coal to gas. However, the report notes that the pause in emissions growth is still not enough to stabilise the climate system at below 2 degrees above pre-industrial temperature levels (we are already one degree above).

In fact, as indicated by the endless, breathless media pronouncements on this point, it would seem that nothing short of completely stopping all industrial activity would satisfying the two degree requirement. As that is obviously not going to happen and with the Paris agreement proving to be completely ineffectual, perhaps it is time to declare international efforts to control emissions a waste of time and move on.

Instead of accepting this point, activists have retreated into fantasies about increases in green energy, including triumphant pronouncements about massive investments in renewable energy in various countries. There you are, a few wind turbines and PV panels will be installed in these countries and they won't need coal. Right!

Green fantasies about the end of the coal industry are beyond the scope of this book, but we will look briefly at the Chinese energy scene. Activists will often declare that about 25 per cent of China's electricity is generated by renewables. This is true, but the activists carefully avoid mentioning that the vast bulk of that renewable energy is from hydro-electricity (which counts as a renewable), and that most of the dams supplying the hydro power were built long before present anti-carbon enthusiasms. The massive Three Gorges project, for example, completed in 2012 is claimed to have the largest power station in terms of installed capacity.

A more telling statistic about China which activists seldom mention, perhaps because they do not know about it, is that the country accounts for about half of the world's coal production. All of that production is used to feed electricity generators and smelters, and still more coal has to be imported from Australia and Indonesia to meet demand. In 2016, the Chinese government deliberately curbed the production of its coal mines. A few hailed this move as the first step to a coal free economy, but analysts noted that the change had more to do with boosting prices, which it did quite effectively, so that still operating mines could operate at a profit rather than cut each other's throats. Coal mines are a major employer in

China. These restrictions were eased to make coal cheaper over the winter.

In all of this, readers will note, climate does not figure. Chinese government decisions are influenced far more by the need to maintain the economy and employment and avoid social unrest than climate. Chinese citizens are right to be very concerned about pollution as the country's cities are very badly polluted, but carbon dioxide is not part of that problem. The problem is due to the likes of aerosols and sulphates (which may be produced in the same emissions as CO_2), combined with third world habits. Search online and you will find stories of poor Chinese families that now cluster in huge numbers around the cities, where the new jobs are, trying to keep warm by burning coal briquettes. Burning what? Anyone above 60 may remember seeing coal briquettes for sale in Australia, but the fires fuelled by such material have long been swept away by the vastly more efficient and less polluting gas and electric heaters, and the air in Australian cities has been the better for it. Large numbers of domestic coal or wood fires are bound to cause a huge smog problem.

The Chinese authorities have reacted in various ways to this issue which need not concern us here, but they have also been sufficiently influenced by Western notions of controlling emissions to introduce a national emissions trading scheme. This scheme was announced in 2015 when China and the then US President Barack Obama struck a bilateral deal over limiting emissions. Nonsense about trading schemes certainly fools Western activists, but it is difficult to see how such a scheme has any relevance to the briquette fires of poor families trying to stay warm.

As we can see the whole exercise, including endless meetings and complex agreements, has been a colossal waste of time. One way to reduce emissions may be to simply ban international meetings on agreements to limit emissions, or insist that delegates can only attend these meetings in person if they can get there on bicycles, without eating meat.

7

What the IPCC now says?

Activists continually repeat earlier IPCC forecasts of doom and gloom outside of temperature forecasts without realising that the panel's fifth assessment report in 2014 toned down claims of climate effects in various areas. That retreat was very likely a result of the panel being caught out using grey material (that is, non-scientific matter) as sources, but as there is very little independent reporting on the panel it is difficult to know for certain Perhaps the sixth assessment report, due out around 2021, will restate the outrageous claims evident in earlier reports, but for now the 2014 report rules.

Health

"In recent decades, climate change has contributed to levels of ill health (likely) though the present worldwide burden of ill health from climate change is relatively small compared with other stressors on health and is not well quantified." – *5ᵗʰ Assessment Report, Impacts Adaptation and Vulnerability. Chapter*

11 Human Health: Impacts, Adaptation, and Co-Benefits.

As we saw in the discussion on Ebola, even that ghastly disease may not have made that much difference compared to all the other diseases which take life in West Africa. Previous IPCC reports forecast major death tolls from malaria, amongst other problems.

Economic effects

"In sum, estimates of the aggregate economic impact of climate change are relatively small but with a large downside risk. Estimates of the incremental damage per tonne of CO_2 emitted vary by two orders of magnitude, with the assumed discount rate the main driver of the differences between estimates. The literature on the impact of climate and climate change on economic growth and development has yet to reach firm conclusions. There is agreement that climate change would slow economic growth, by a little according to some studies and by a lot according to other studies. Different economies will be affected differently. Some studies suggest that climate change may trap more people in poverty." *5th Assessment Report Impacts, Adaptation and Vulnerability. Chapter 10 Key Economic Sectors and Services.*

This is quite different to previous claims of the vast economic effects that will occur from climate change, including the 2006 Stern Review, named after the economist who orchestrated it,

Nicholas Stern. By far the most strident of the independent reports, this concluded that strong, early action on stabilising carbon dioxide levels was justified in order to avoid the economic damage higher temperatures would cause. The discount rate referred to is something like a long-term real return on investment which economists use to decide whether it is worth spending a dollar now to offset so many dollars-worth of damage in the future. Stern set the rate at a very low one per cent or so but as the IPCC correctly points out, there is a wide range of opinions on this point. In any case, as the Stern report was a decade ago, the opportunity for early action option - never feasible anyway – has passed.

Major storms

"Globally there is low confidence regarding changes in tropical cyclone activity over the 20th century owing to changes in observational capabilities, although it is virtually certain that there has been an increase in the frequency and intensity of the strongest tropical cyclones in the North Atlantic since the 1970s (WGI AR5 Section 2.6). In the future, it is likely that the frequency of tropical cyclones globally will either decrease or remain unchanged, but there will be a likely increase in global mean tropical cyclone precipitation rates and maximum wind speed (WGI AR5 Section 14.6)." – *5th Assessment Report, Impacts, Adaptations and Vulnerabilities. Chapter 5, Coastal Systems and Low Lying Areas.*

This careful, cautious statement is a world away from the doom mongering of activists. In the future storms may get a little worse.

Sea levels

The worst estimate for sea level increases given in the same chapter as the one cited above is for just under a metre, as a possible variation on a mean sea level increase of 0.74 metres. As noted in the chapter on farcical fears, activists have not been happy with that estimated increase as it is not dramatic enough. The US National Oceanic and Atmospheric Administration has raised the bidding in this area with modelling which puts the rise at two metres or so by 2100. Activists think that's a major improvement, but would like an extra metre or so. No doubt someone will oblige.

While on the subject of sea levels the IPCC makes this interesting admission.

It is virtually certain that globally averaged sea level has risen over the 20th century, with a very likely mean rate between 1900 and 2010 of 1.7 [1.5 to 1.9] mm yr−1 and 3.2 [2.8 and 3.6] mm yr−1 between 1993 and 2010. This assessment is based on high agreement among multiple studies using different methods, and from independent observing systems (tide gauges and altimetry) since 1993. It is likely that a rate comparable to that since 1993 occurred between 1920 and 1950, possibly due to a multi-decadal climate variation, as individual tide gauges around the world and all reconstructions of GMSL (global mean sea level) show increased rates of sea level rise during this period. - *Working Group I of the IPCC's Fifth Assessment Report: Chapter 3 – Observations: Oceans.*

In other words, the report is pointing to an overall, average increase for the 20th century and a higher one, as measured by satellite between 1993 and 2010, but it also admits that sea levels probably rose at a rate comparable to that of the satellite measurements between 1920 and 1950. This would seem to counter claims that the increase in sea levels are accelerating but, as noted in the chapter on *Farcical Fears*, we do not need to trouble ourselves about this area. We can wait to see what happens and react to the actual results.

Conclusion: why bother?

In a story in the *Australian Financial Review* in early January 2019 one green industry executive who we shall not bother to name here, complaining about the lack of action about climate matters in the Liberal Party, stating that: "actual climate events are occurring with regularity".

What was the executive referring to? The extreme cold and series of avalanches in Europe at that time? There had been hot days and storms in Australia's South East in the preceding weeks, but those are hardly surprising in the region in January. Parts of the South East were drought stricken, but there is nothing new about droughts in Australia. Tropical cyclones are the most spectacular of the local weather/climate events, but none had crossed the coast in the 2018-19 season (November to April) to that point although at least one is expected per season. The Northern Hemisphere's hurricane season (May to November) had admittedly been more severe than usual with Hurricane Michael killing people and damaging homes in several countries but, again, it is difficult to point to anything unusual or unprecedented in even that event.

In other words, as noted at the beginning of this essay, those involved in the global warming business are convinced that doom is just around the corner and never mind actual events. But for those of us who live in the real world where are the big temperature increases and unprecedented weather events that we have been repeatedly promised over more than 30 years of screaming and shouting about emissions? Instead we are told that the failure of computer models to forecast the fairly modest temperature increases that have occurred is simply due to unforeseen variations. The big changes will be along shortly, don't you worry about that! In the meantime, variations in the weather are co-opted to prove the point. It's unusually hot today so therefore climate change must be happening, and never mind the fact that its January.

Activists who accept those arguments also brush aside the evident problem that international efforts to curb emissions have failed miserably. Activists campaign as if future climate hinges on Australia's effort to reduce emissions when, in fact, if the science is right, those efforts are at best a gesture – an effort to be seen to be doing something. Whether the community at large wants to make that gesture is a matter for public debate, but the issue should never be presented as one where Australia can affect emissions by itself. This is not possible.

An often used backup argument is that Australia should show moral leadership in this area by aggressively cutting emissions. It is difficult to see countries like China, Russia or Turkey paying much attention to Saint Australia in this regard, especially with the US going its own way, but again that is a matter for the community after they have been fully informed that we are spending money and making real sacrifices to show moral leadership.

That same community might well be asked that if we are going to spending money on climate issues is it better to spend it trying to curb emissions, which will never have any effect, or on area such as changing building codes, so that the structures (and the people inside them) are far less vulnerable to damage from storms or fires? Instead of building more wind farms why not pay for more flood control works upstream of the flood plain of the Brisbane river? If we are going to spend money, why not spend it on projects that we know will make a difference? This sort of pragmatic approach has been popularised by Danish academic Dr Bjorn Lomborg, who has been pilloried by the global warming industry for his troubles. This common sense approach is not what activists want to hear.

None the less those same academics have yet to come up with any satisfactory refutation of Lomborg's approach. Instead they want to continue with efforts to limit carbon dioxide emissions, the sole result which to date has been to give jobs and all-expenses paid overseas trips to activists.

www.ingramcontent.com/pod-product-compliance
Lightning Source LLC
Chambersburg PA
CBHW071114210326

41519CB00020B/6293